Health Benefits of Alpha Lipoic Acid

Exploring the Science of
Alpha Lipoic Acid Supplements

By John Iovine

Copyright

All Rights Reserved. No part of this book may be reproduced in any manner without written permission from the publisher. While every precaution has been taken in the preparation of this book, the author, the publisher, or the seller assumes no responsibility for errors or omissions. No liability is assumed for damages resulting from the use or misuse of the information contained herein.

READ THIS FIRST – DISCLAIMER
The author has narrated his research experiences in this book by observing and evaluating facts and figures. The reliance on information has been done in good faith and is believed to be reliable, according to the author's best knowledge. The sources of referenced information could change or be updated in the future. The author cannot guarantee the validity and accuracy of the sources, which may change, be modified, updated, or removed in the future, and thus, disclaims himself from any such changes, modifications, updates, and removals. (3)

ISBN: 978-1-62385-020-3 Print
ISBN: 978-1-62385-019-7 eBook

Table of Contents

1 - What is Alpha Lipoic Acid? 3
2 - Brief History of Alpha Lipoic Acid 5
3 - Benefits of Alpha Lipoic Acid 6
4 - Supplemental Alpha Lipoic Acid Isn't a Fad 6
    Synergy (ALA + RLT + H2) 7
5 - Alpha Lipoic Acid's Antioxidant Properties 7
6 - Alpha Lipoic Acid is a Potent Anti-Inflammatory 9
7 - Neuroprotective Effects of Alpha Lipoic Acid 10
8 - ALA's Role in Cognitive Function and Memory 11
9 - Alpha Lipoic Acid and Blood Sugar Regulation 12
10 - Alpha Lipoic Acid lowers Blood Pressure 13
11 - Alpha Lipoic Acid for Weight Loss 14
12 - Alpha Lipoic Acid for Treating Eye Disease 14
13 - Alpha Lipoic Acid for Cardiovascular Health 15
14 - Alpha Lipoic Acid for Heavy Metal Chelation 16
15 - Alpha Lipoic Acid for Enhanced Athletic Performance 16
16 - Reduction of Oxidative Stress during Exercise 17
17 - The Aging Process and Alpha Lipoic Acid 19
18 - Adding Alpha Lipoic Acid into Your Daily Routine 19
19 - Recommended Dosage of Alpha Lipoic Acid 20
20 - Selecting the Best Alpha Lipoic Acid Supplement 21
21 - Potential Side Effects and Precautions 22
Appendix: 24
Bibliography 25

1 - What is Alpha Lipoic Acid?

Alpha Lipoic Acid (ALA) is a powerful antioxidant that is naturally produced in the body and assists in converting glucose into energy. ALA, which is found naturally in foods, such as carrots, spinach, tomatoes, and broccoli, is the natural R-isomer version. The small quantities available in food make extraction for vitamin supplements impractical. ALA supplements on the market are manufactured and synthetic compounds.

It is necessary to take dietary supplements to achieve the dosages used to obtain the benefits ascribed in the studies quoted.

ALA is a regulated drug in a few European countries. Fortunately, ALA is available in the United States as a dietary supplement.

ALA is a unique antioxidant that is both water and fat-soluble. This allows the body to utilize ALA in neutralizing harmful free radicals that can cause oxidative stress and damage cells. By combating oxidative stress, ALA helps protect against various chronic diseases, including heart disease, diabetes, and certain cancers.

ALA is becoming increasingly popular among people, athletes, and seniors due to its potential to improve overall health and well-being.

ALA supports healthy aging. As we age, our cells are exposed to increased oxidative stress, which can lead to accelerated aging and age-related diseases. ALA helps combat this by enhancing the activity of other antioxidants, such as vitamins C and E, and promoting your body's production of glutathione, a key antioxidant enzyme.

ALA has been shown to modestly improve insulin sensitivity and blood glucose control. This is beneficial to people with diabetes, per-diabetics, and metabolic syndrome.

For people and athletes, improved insulin sensitivity can enhance glucose uptake by muscle cells, leading to improved energy production and better athletic performance. ALA also helps reduce inflammation and muscle damage caused by intense exercise, aiding post-workout recovery. Additionally, ALA has been found to increase the production of adenosine triphosphate (ATP), the primary energy source for cells, which can boost endurance and stamina during physical activity.

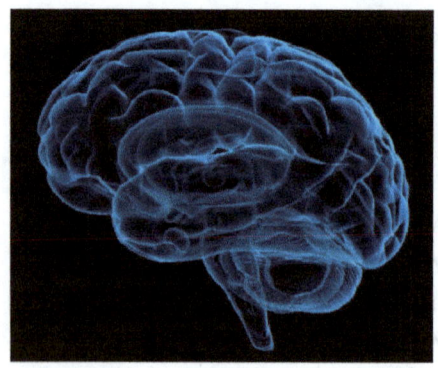

Studies on seniors have shown that ALA can improve cognitive function and memory, making it a promising tool in preventing and managing age-related cognitive decline, such as Alzheimer's disease. Furthermore, ALA has been found to protect against diabetic neuropathy, a common complication of diabetes that affects the nerves and can lead to pain, numbness, and tingling in the extremities.

2 - Brief History of Alpha Lipoic Acid

Alpha lipoic acid was initially identified as a component of a larger biological molecule in the 1930s by enzyme researchers. In 1951, ALA was isolated by Lester J. Reed and Irwin C. Gunsalus. The compound is found in the mitochondria of cells and is identified as a component of cellular metabolism. However, its first clinical use wasn't until 1959. Scientists discovered its unique antioxidant properties. They found that ALA could neutralize harmful free radicals in the body, thereby protecting cells from oxidative damage.

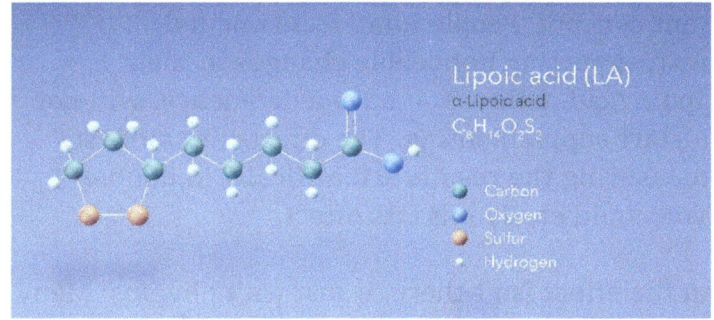

Further research in the 1980s and 1990s revealed that ALA plays a crucial role in energy production. It was found to be a cofactor for enzymes involved in converting glucose into usable energy. This discovery led to its use as a dietary supplement to support energy levels and metabolic function.

3 - Benefits of Alpha Lipoic Acid

- Improves Weight Loss
- Reduces Oxidative Stress in the Body
- Improves Insulin Sensitivity
- Improves Fasting Blood Glucose
- Reduces Blood Pressure
- Removes Toxic Metals from the body.
- Improves Diabetic Neuropathy
- Offers Neuroprotection
- Improves visual function.
- Reduces Insulin Resistance
- Reduces Brain Damage due to Stroke.

4 - Supplemental Alpha Lipoic Acid Isn't a Fad

Alpha lipoic acid is a potent broad-spectrum antioxidant that benefits people of all ages, athletes, and seniors. It is important to keep in mind that the effects of ALA are scientifically measurable, as shown in the studies, but modest.

Modest is an important concept. People jump from one fad supplement to the next, expecting noticeable changes in their physiology. Alpha Lipoic Acid, unless you have a deficiency, will not ring a bell when you start supplementing. The effects are modest, and unless you are measuring the results scientifically in double-blind experiments, you could easily miss its effect.

Modest doesn't mean the effect isn't there, it just isn't obvious. Many leading-edge therapies also have modest effects.

Synergy (ALA + RLT + H2)

Alpha Lipoic Acid is one of three current treatments that produce positive, modest, but measurable scientific results. The second treatment option is Red Light Therapy (RLT), and the third is hydrogen therapy. These therapies also boost aging mitochondria. Hydrogen therapy, like ALA, is a powerful antioxidant.

The question is, if a person took ALA supplements, combined with hydrogen therapy and RLT, would the combination create a potent synergic treatment option?

This is a question I hope is answered in the near future.

5 - Alpha Lipoic Acid's Antioxidant Properties

Antioxidants play a crucial role in safeguarding our bodies against the damaging effects of free radicals. Free radicals are unstable molecules that can cause oxidative stress, leading to cell damage and various health issues. ALA acts as a potent antioxidant by neutralizing free radicals, helping to protect our cells and tissues from harm.[1]

OXIDATIVE STRESS

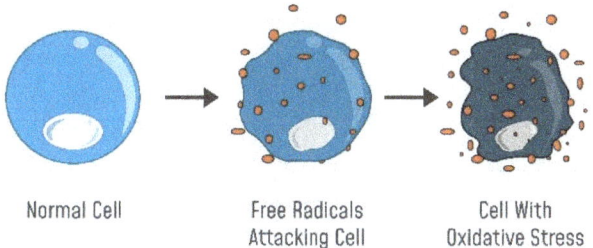

Normal Cell | Free Radicals Attacking Cell | Cell With Oxidative Stress

Since ALA is a fat and water-soluble antioxidant, it can work throughout the body, including the brain, nerves, and liver, to combat oxidative stress.

Role of Alpha Lipoic Acid in Energy Production

Alpha Lipoic Acid (ALA) plays a crucial role in energy production within the body. ALA is naturally synthesized inside the mitochondria of cells, acting as a cofactor for energy metabolism.[2]

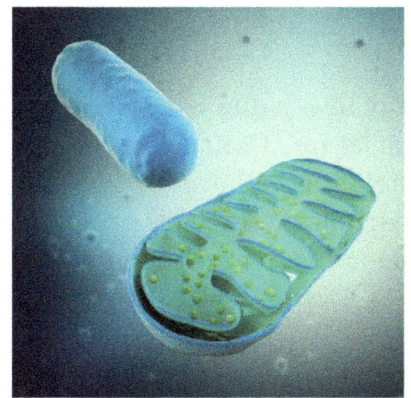

ALA is involved in the mitochondria's production of adenosine triphosphate (ATP), the energy currency of our cells. ATP provides energy for various cellular processes, including muscle contraction, nerve signaling, and metabolism. ALA acts as a cofactor for critical enzymes that convert glucose into ATP, ensuring a steady energy supply.

Adenosine triphospate (ATP)

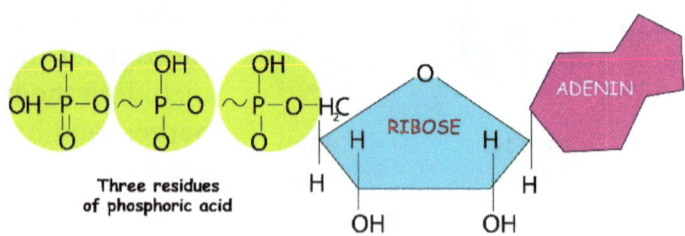

For athletes, ALA can be particularly beneficial due to its ability to enhance glucose uptake by muscle cells. By increasing glucose transporters on the cell membrane, ALA facilitates glucose uptake from the bloodstream into the muscles, which can be used as a fuel source during exercise. This can lead to improved performance, endurance, and quicker recovery times.

Seniors often experience a decline in energy levels due to age-related factors; one factor is decreased mitochondrial function. Mitochondria are the powerhouses of our cells, responsible for ATP production. ALA has been shown to enhance mitochondrial function and protect them from oxidative damage, ultimately boosting energy levels in seniors.

6 - Alpha Lipoic Acid is a Potent Anti-Inflammatory

Alpha lipoic acid (ALA) can effectively combat inflammation and its associated health conditions.

Inflammation is a natural bodily response to injury or infection. However, when inflammation becomes chronic, it can contribute to the development of heart disease, diabetes, and even Alzheimer's. As a potent antioxidant, ALA reduces oxidative stress, helping to quell inflammation and promote healing.

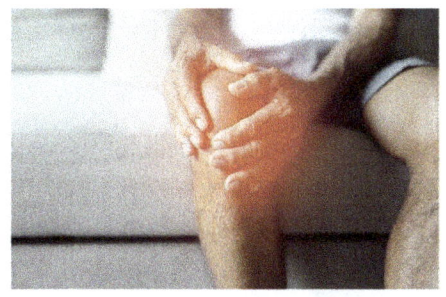

For athletes, ALA can assist in managing exercise-induced inflammation, aiding in faster recovery, and reducing muscle soreness.

For seniors, ALA can reduce chronic inflammation due to aging or age-related health conditions.

7 - Neuroprotective Effects of Alpha Lipoic Acid

The brain is highly susceptible to oxidative stress. ALA helps reduce oxidative damage and protects the brain from neurodegenerative diseases such as Alzheimer's and Parkinson's.

Alpha lipoic acid has been shown to enhance cognitive function and memory. Multiple studies have demonstrated that ALA supplementation can improve learning abilities, concentration, and overall mental performance.

Nerve regeneration. Neurological conditions such as peripheral neuropathy, often experienced by athletes and seniors, can result in nerve damage and loss of sensation. ALA has been found to stimulate the regeneration of damaged nerves, leading to improved nerve function and reduced symptoms. This makes it a valuable supplement for athletes recovering from injuries and seniors battling age-related nerve damage.

Some research shows ALA can counteract nerve damage caused by chemotherapy.[3] In addition, improve the quality of life for cancer patients.[4]

Alpha lipoic acid holds significant potential as a neuroprotective supplement. Its antioxidant, cognitive-enhancing, nerve-regenerating, and anti-inflammatory properties make it a beneficial addition to the lives of people, athletes, and seniors.[5,6]

8 - ALA's Role in Cognitive Function and Memory

ALA has been shown to enhance memory and learning abilities. Studies have demonstrated that ALA supplementation can improve memory recall, information retention, and overall cognitive function. This is particularly beneficial for students, professionals, and anyone seeking to enhance their mental acuity.

For seniors, ALA's role in cognitive function and memory becomes even more crucial. Age-related inflammation is associated with a decline in cognitive function and an increased risk of neurodegenerative diseases such as Alzheimer's. ALA has been shown to cross the blood-brain barrier, where it can protect brain cells from oxidative damage and reduce inflammation.[7,8]

Alzheimer's Disease

There is a small amount of promising research that suggests ALA may be beneficial for treating neurological diseases like Alzheimer's and Parkinson's disease. This is an area of research, study, and debate.[9]

9 - Alpha Lipoic Acid and Blood Sugar Regulation

Alpha lipoic acid (ALA) has gained significant attention in blood sugar regulation for those struggling with diabetes.[10,11]

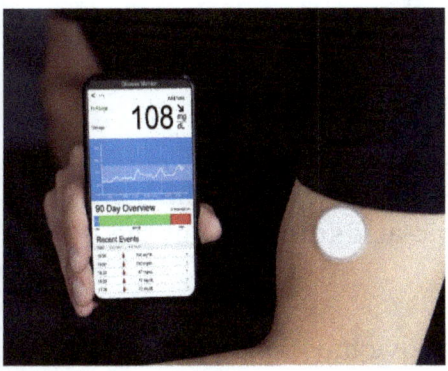

ALA taken at 300 to 1200 mg daily modestly enhances insulin sensitivity. Insulin is a hormone responsible for regulating blood sugar levels by facilitating the uptake of glucose into cells. However, in conditions like diabetes, insulin resistance occurs, leading to elevated blood sugar levels. ALA has been shown to modestly increase insulin efficiency, allowing cells to respond better to its signals and promoting better blood sugar control.[12]

For seniors, ALA's modest impact on glucose metabolism can help prevent age-related conditions such as type 2 diabetes. By maintaining healthy blood sugar levels, ALA can reduce the risk of developing insulin resistance and metabolic disorders commonly associated with aging.

Diabetic neuropathy is nerve damage to the body caused by the toxic effects of high blood glucose. This nerve damage is characterized by numbness and sensations of burning, tingling, itching, and pain.

Diabetic Neuropathy

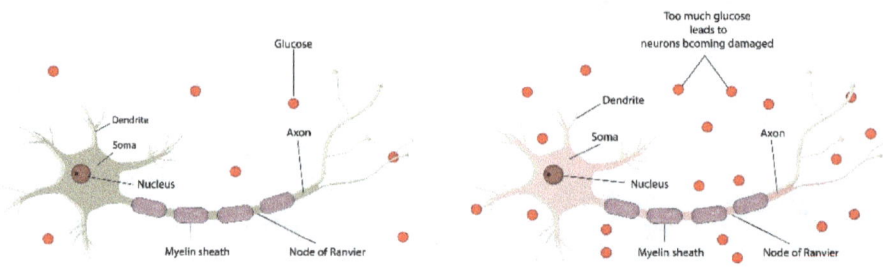

Placebo-controlled studies show 600-1,800 mg of ALA daily significantly improved diabetic neuropathy. Because of the high dosage required, using the R-alpha-lipoic-acid form can cut the dosage in half.[13]

For athletes, ALA has been found to improve muscle glucose uptake, aiding in post-workout recovery and promoting muscle growth.

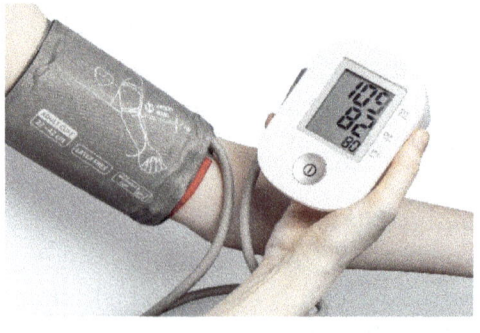

10 - Alpha Lipoic Acid lowers Blood Pressure

Individual studies have given inconsistent results regarding ALA's ability to lower blood pressure. A meta-analysis where ten studies that included a total of 610 participants met the experimenter criteria did show that ALA provided a lowering of blood pressure.

11 - Alpha Lipoic Acid for Weight Loss

ALA supports modest weight loss and may assist by curbing the appetite, boosting energy levels, and enhancing metabolism by stabilizing insulin levels. One study had n=81 overweight men and women supplemented with 600 mg of R-alpha-lipoic acid daily for 24 weeks. All adults began with a BMI of greater than 25 kg/m^2.

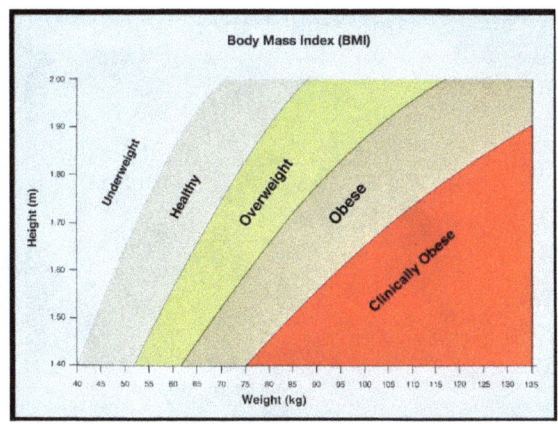

Results: After 24 weeks, the average BMI of the treated group decreased by 6 km/m^2. At the same time, the placebo group's BMI increased by 2 kg/m^2.

This means that the modest weight loss of fat was supplemented with a modest increase in muscle tissue. A positive change in their overall body composition. All participants were asked not to change their eating or exercise habits during the study, which leads me to believe these participants lost fat and gained muscle tissue without diet or exercise.

Studies show that overweight people who are put on a calorie-restrictive diet lose more weight if they supplement with 300 mg of ALA.[14,15,16]

12 - Alpha Lipoic Acid for Treating Eye Disease

Cataracts - As adults age, their susceptibility to cataracts increases, leading to clouded vision due to lens opacities. A significant factor contributing to cataract formation is oxidative stress within the eye's lens. In experimental animal studies, alpha lipoic acid has shown promise in preventing cataracts. Researchers theorize that lipoic acid

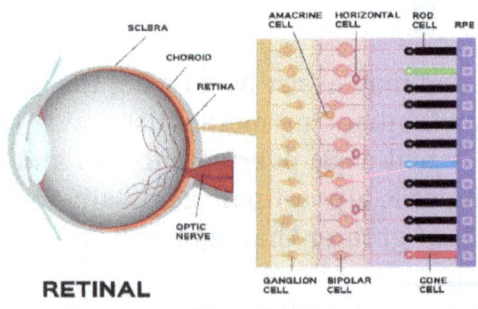

provides this protection by enhancing the levels of crucial antioxidant enzymes, such as glutathione peroxidase.

Glaucoma - A study involving patients with open-angle glaucoma demonstrated improved visual function and other glaucoma indicators in participants who either took 75 mg of lipoic acid daily for two months or 150 mg daily for one month. This was compared to a control group that received no lipoic acid. Significant improvement was seen in about 50% of the eye cases.[17]

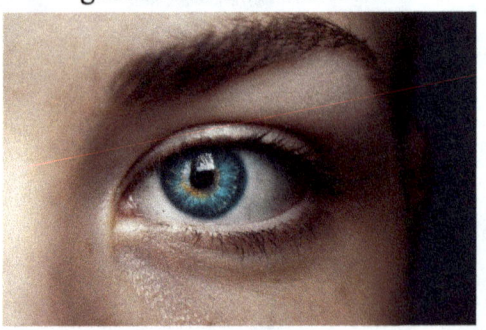

Retinitis Pigmentosa - A combination of anti-oxidants that include alpha lipoic acid, vitamin E, and vitamin C can potentially prevent retinal cell death in animals suffering from retinitis pigmentosa. When this antioxidant mixture was injected daily into the eyes of mice, it helped preserve the retina cells. This condition also impacts humans. Given that there isn't a medical solution for this debilitating disease yet, the potential of a nutritional intervention offers an exciting and hopeful prospect.[18]

New studies suggest that ALA slows the progress of macular degeneration.[19]

13 - Alpha Lipoic Acid for Cardiovascular Health

ALA helps prevent the oxidation of cholesterol, reducing the risk of plaque formation in the arteries. This, in turn, enhances blood flow and lowers the chances of developing conditions such as atherosclerosis, heart attacks, and strokes.

Alpha lipoic acid is an impressive ally for cardiovascular health. Its antioxidant properties, ability to improve blood flow, and reduce inflammation make it an ideal supplement for people, athletes, and seniors.

14 - Alpha Lipoic Acid for Heavy Metal Chelation

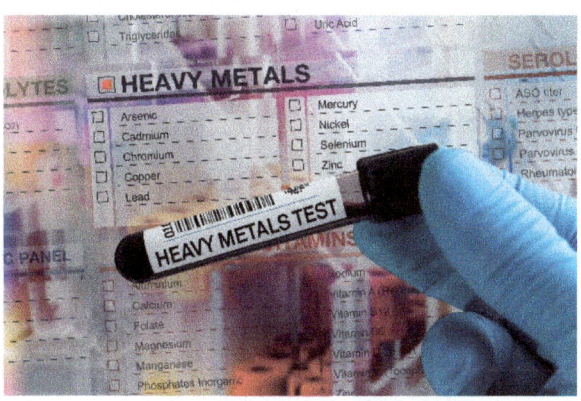

Alpha lipoic acid can help the body eliminate toxic metal contaminants typically found in food and the environment. ALA renders arsenic, cadmium, lead, and mercury inactive so they can be removed from the body. Using rats, animal studies have shown that ALA can remove toxic metals from biological systems.[20,21]

15 - Alpha Lipoic Acid for Enhanced Athletic Performance

ALA enhances your body's ability to produce and utilize energy efficiently. By supporting energy production at the cellular level, ALA allows you to stay energized and productive.

Senior's energy levels naturally decline, making it harder to stay active and engaged. ALA can help counteract this decline by boosting mitochondrial function.

By supplementing with ALA, seniors can experience increased vitality and maintain a higher quality of life.

16 - Reduction of Oxidative Stress during Exercise

Exercise is an essential part of maintaining a healthy lifestyle. Physical activity can lead to an increase in oxidative stress within the body.

By supplementing with alpha lipoic acid, athletes and individuals engaging in regular exercise can improve athletic performance and enhance recovery.

Enhanced Muscle Recovery with Alpha Lipoic Acid

ALA has been shown to accelerate muscle recovery, allowing athletes to bounce back faster and perform at their peak. Alpha Lipoic Acid is a powerful tool for enhancing muscle recovery in people of all ages and athletic abilities.[22]

Alpha Lipoic Acid's Impact on Exercise-Induced Inflammation

For athletes, inflammation is a common occurrence due to intense physical activity. ALA can assist in managing exercise-induced inflammation, aiding in faster recovery, and reducing muscle soreness. Its ability to reduce inflammation makes it an attractive option for seniors who may experience chronic inflammation due to aging or age-related health conditions.

ALA helps to reduce the levels of pro-inflammatory cytokines in the body. By inhibiting the production of these inflammatory molecules, ALA can help to minimize the damage caused by exercise-induced inflammation.

17 - The Aging Process and Alpha Lipoic Acid

Counteracting Age-Related Oxidative Stress

ALA can contribute to optimal health and well-being by counteracting age-related oxidative stress and age-related inflammation.

Chronic inflammation can damage blood vessels and lead to atherosclerosis, increasing the risk of heart disease and stroke. ALA has been shown to improve endothelial function, reduce inflammation markers, and lower blood pressure. By managing age-related inflammation with ALA, individuals can promote a healthy heart and reduce the risk of cardiovascular diseases.

Alpha Lipoic Acid for Healthy Aging and Longevity

Not only does alpha lipoic acid possess antioxidant properties, but it also plays a vital role in regenerating other antioxidants in the body, such as vitamins C and E, glutathione, and coenzyme Q10. This unique ability allows ALA to enhance the antioxidant defense system and amplify its protective effects.

18 - Adding Alpha Lipoic Acid into Your Daily Routine

It is essential to consult with a healthcare professional before starting any new supplementation regimen, as individual needs may vary.

There isn't a RDA, Recommended Daily Dose for ALA. Individuals with kidney, liver, thyroid, diabetes, or pregnant and nursing women should exercise caution and speak to a healthcare provider before supplementing.

19 - Recommended Dosage of Alpha Lipoic Acid

Most of the studies were performed with mixed isomer ALA. You can cut the dosage in half if one purchases the more expensive R-alpha-lipoic acid. This may be a consideration if taking ALA is causing any heartburn or digestive distress.

The ALA can be split into three equal doses during the day. So, for a 300 mg daily dose, you can take 100 mg tablets 3X daily.

If tolerated, ALA should be taken on an empty stomach. If taken with food, you may lose 30% of the absorption of the ALA dose.

A daily dosage of 300-600 mg of ALA is often recommended for general health maintenance. This dosage can help combat oxidative stress, enhance cellular energy production, lose weight, and support overall well-being.

Weight Loss: 100-200 mg tablets taken three times a day with meals.

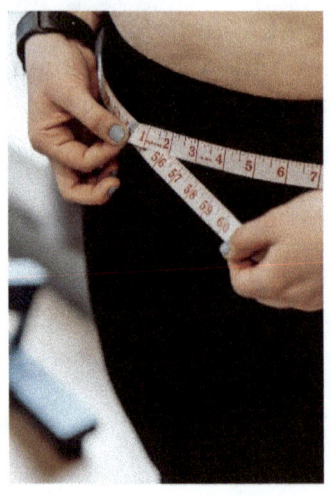

For athletes and individuals engaged in intense physical activities, a slightly higher dosage of ALA may be beneficial, 600-900 mg, to help reduce oxidative damage caused by rigorous exercise, boost muscle recovery, and enhance performance. Consider dividing the dosage throughout the day to maintain a consistent level of ALA in their system and or using R-Alpha Lipoic Acid to cut the dosage in half.

Seniors, who often experience age-related decline in cellular function and increased oxidative stress, can also benefit from ALA supplementation. A 600-900 mg dosage per day has shown promising results in improving cognitive function, reducing inflammation, and supporting overall brain health. Additionally, ALA can help manage blood sugar levels and protect against diabetic complications, which are common concerns among seniors.

20 - Selecting the Best Alpha Lipoic Acid Supplement

It is essential to check the purity and quality of the alpha lipoic acid supplement. Look for a reputable brand that follows Current Good Manufacturing Practices (CGMP) as defined by the FDA. It will have CGMP or GMP stamped on the vitamin container.

One important consideration is the bioavailability of the ALA. Bioavailability refers to the extent and rate at which a substance is absorbed into the bloodstream and reaches its target tissues.

ALA is not extracted from natural sources but are manufactured. As such, the ALA supplements contain two synthetic isomers, or mirror images of the ALA compound. These mirror compounds are identified as the R- and S-isomers. Only the R-isomer has been shown to be biologically active. The ALA supplements you purchase in a vitamin store contain 50% S-isomers.

ALA supplements with just the R-Alpha Lipoic Acid are available but are more expensive. These supplements are identified as R-Alpha Lipoic Acid and have the S-isomer compound removed. You can reduce the recommended dosage in half if you purchase the more expensive R-Isomer supplement.

21 -Potential Side Effects and Precautions

1. Allergic Reactions: Although rare, some individuals may experience allergic reactions to alpha lipoic acid. This can manifest as skin rashes, itching, or swelling. If you have a known allergy to ALA or any related compounds, it is advisable to avoid its use or consult a healthcare professional before taking it.

2. Hypoglycemia: Alpha lipoic acid can potentially lower blood sugar levels. While this can benefit individuals with metabolic syndrome or prediabetes, adjustments may be necessary for people taking anti-diabetic medication.

3. Interactions with Medications: Alpha lipoic acid may interact with certain medications, such as chemotherapy drugs, thyroid medications, and medications for diabetes. It is crucial to inform your healthcare provider about any supplements you are taking to prevent any potential interactions or adverse effects.

4. Gastrointestinal Upset: Alpha lipoic acid is an acid (a bit stronger than acetic acid in vinegar) and may cause mild gastrointestinal symptoms such as nausea, stomach cramps, or diarrhea. To minimize these side effects, it is recommended to take ALA with food or divide the dosage throughout the day.

5. ALA may cause a strong order in urine due to sulfur content.

6. In Patients with thyroid disease, ALA may interfere with the conversion of T4 (thyroxine) to T3 (triodothyronine), lowering T3 levels.

Alpha lipoic acid's potential benefits are vast and far-reaching. From improving overall health to enhancing athletic performance and promoting healthy aging, the future of ALA research is poised to unlock even more remarkable discoveries. By staying informed and incorporating this powerful antioxidant into our daily routines, we can harness its potential for optimal health and well-being.

If you like this book please leave a review on Amazon.
https://www.amazon.com/review/create-review?&asin=B0CH5PHXW1

To send me a comment on this book, use my email address below.
johns-books@proton.me

More health books by John Iovine

Red Light Therapy
https://www.amazon.com/dp/B09FC59M4Z

Understanding Fat - The Secrets To Losing Weight
https://www.amazon.com/dp/B09NKDPXPG

Hydrogen For Health
https://www.amazon.com/dp/B0CN8MPL1R

Appendix:
Glossary of Terms

1. Alpha Lipoic Acid (ALA) is known for regenerating other antioxidants, such as vitamins C and E, and plays a crucial role in energy production.

2. Antioxidant: Substances that protect our cells from damage caused by harmful molecules called free radicals. Antioxidants help neutralize these free radicals and reduce oxidative stress associated with various health conditions and aging.

3. Oxidative Stress: An imbalance between the production of free radicals and the body's ability to counteract their harmful effects. Oxidative stress can lead to cellular damage and is believed to contribute to the development of chronic diseases.

4. Mitochondria: Often referred to as the powerhouses of our cells, mitochondria produce the energy needed for cellular function. Alpha lipoic acid plays a crucial role in supporting mitochondrial health and function.

5. Insulin Sensitivity: The ability of our cells to respond effectively to insulin, a hormone that regulates blood sugar levels. Alpha lipoic acid has been shown to improve insulin sensitivity, making it beneficial for individuals with diabetes or insulin resistance.

6. Neuropathy: A condition characterized by nerve damage, leading to symptoms such as pain, numbness, and tingling. Alpha lipoic acid has been studied for its potential to alleviate neuropathic symptoms, particularly in individuals with diabetic neuropathy.

7. Inflammation: The body's natural response to injury or infection, characterized by redness, swelling, and pain. Chronic inflammation is associated with various health conditions, and alpha lipoic acid has been shown to possess anti-inflammatory properties.

8. Cognitive Function: The ability to think, learn, and remember. Research suggests that alpha lipoic acid may positively impact cognitive function, making it relevant for individuals of all ages, particularly seniors.

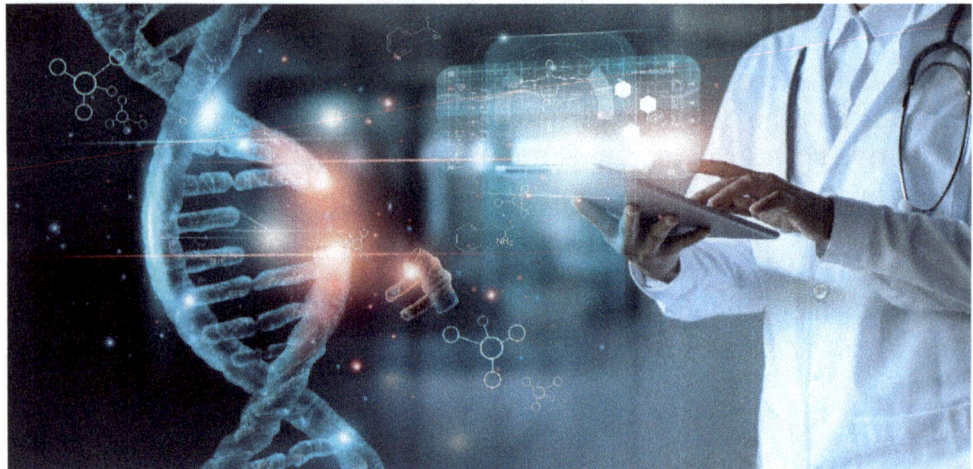

Bibliography

1 - Antioxidant and anti-inflammatory properties of alpha-lipoic acid protect against valproic acid-induced liver injury
https://pubmed.ncbi.nlm.nih.gov/33275538/

2 - Alpha-lipoic acid as a dietary supplement: molecular mechanisms and therapeutic potential
https://pubmed.ncbi.nlm.nih.gov/19664690/

3 - Oral alpha-lipoic acid to prevent chemotherapy-induced peripheral neuropathy: a randomized, double-blind, placebo-

controlled trial
https://pubmed.ncbi.nlm.nih.gov/24362907/

4 - Alpha Lipoic Acid and Cancer
https://www.cancertherapyadvisor.com/home/tools/fact-sheets/alpha-lipoic-acid-and-cancer/

5 - Lipoic acid as an anti-inflammatory and neuroprotective treatment for Alzheimer's disease
https://pubmed.ncbi.nlm.nih.gov/18655815/

6 - Lipoic acid as a novel treatment for Alzheimer's disease and related dementias
https://pubmed.ncbi.nlm.nih.gov/16989905/

7 - Potential Therapeutic Effects of Lipoic Acid on Memory Deficits Related to Aging and Neurodegeneration
https://www.ncbi.nlm.nih.gov/pmc/articles/PMC5732919/

8 - Alpha lipoic acid treatment in late middle age improves cognitive function: Proteomic analysis of the protective mechanisms in the hippocampus
https://pubmed.ncbi.nlm.nih.gov/36708754/

9 - Decrypting the potential role of α-lipoic acid in Alzheimer's disease
https://www.sciencedirect.com/science/article/abs/pii/S0024320521008869

10 - Effect of alpha-lipoic acid on blood glucose, insulin resistance and glutathione peroxidase of type 2 diabetic patients
https://pubmed.ncbi.nlm.nih.gov/21666939/

11 - Safety and Efficacy of Alpha Lipoic Acid During 4 Years of Observation: A Retrospective, Clinical Trial in Healthy Subjects in Primary Prevention
https://pubmed.ncbi.nlm.nih.gov/33299302/

12 - Improvement of insulin sensitivity in patients with type 2 diabetes mellitus after oral administration of alpha-lipoic acid
https://pubmed.ncbi.nlm.nih.gov/17178700/

13 - Treatment of symptomatic diabetic peripheral neuropathy with the anti-oxidant alpha-lipoic acid. A 3-week multicentre randomized controlled trial (ALADIN Study)
https://pubmed.ncbi.nlm.nih.gov/8786016/

14 - A Randomized Controlled Trial of Long-Term (R)-α-Lipoic Acid Supplementation Promotes Weight Loss in Overweight or Obese Adults without Altering Baseline Elevated Plasma Triglyceride Concentrations
https://pubmed.ncbi.nlm.nih.gov/32692358/

15 - Alpha-lipoic acid (ALA) as a supplementation for weight loss: results from a meta-analysis of randomized controlled trials
https://pubmed.ncbi.nlm.nih.gov/28295905/

16 - Alpha-lipoic acid reduces body weight and regulates triglycerides in obese patients with diabetes mellitus
https://pubmed.ncbi.nlm.nih.gov/26276648/

17 - [Lipoic acid as a means of metabolic therapy of open-angle glaucoma]
https://pubmed.ncbi.nlm.nih.gov/8604540/

18 - Antioxidants reduce cone cell death in a model of retinitis pigmentosa
https://pubmed.ncbi.nlm.nih.gov/16849425/

19 - α-Lipoic Acid Treatment Improves Vision-Related Quality of Life in Patients with Dry Age-Related Macular Degeneration
https://pubmed.ncbi.nlm.nih.gov/27840374/

20 - Chelation: Harnessing and Enhancing Heavy Metal Detoxification—A Review
https://www.ncbi.nlm.nih.gov/pmc/articles/PMC3654245/

21 - ALA as a Heavy Metal Detoxifier—the Pros & the Cons
https://www.wellness.com/blog/13290636/ala-as-a-heavy-metal-detoxifier-the-pros-the-cons/fred-fletcher

22 - The effects of alpha lipoic acid on muscle strength recovery after a single and a short-term chronic supplementation - a study in healthy well-trained individuals after intensive resistance and endurance training
https://pubmed.ncbi.nlm.nih.gov/33261642/

www.ingramcontent.com/pod-product-compliance
Lightning Source LLC
Chambersburg PA
CBHW070038040426
42333CB00040B/1719